# WORKBOOK EDITION

By, Jennie von Eggers
& Marillee Flanagan

*Cover & Interior Artwork Design by Jennie von Eggers*
*Edited by Marillee Flanagan & Janet Haddock*
*Published by Trigger Memory Co. LLC*
*ISBN: 978-0-9863000-4-2*

© Copyright 2017 All rights reserved
Published by Trigger Memory Co. LLC

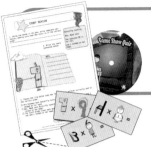

# Contents

WORKBOOK EDITION

**Instructions - How To Use the Workbook**     4
A Peek Inside     5-6

**Meet the Number Symbol Characters**     9-10

**Let's Begin! Part 1 - Upper 3's & 4's**
Learn the Eight Stories     11-26
Crossword Puzzle Challenge     27
Story Review Quiz     28
Light Bulb Moment! How it Works!     29
Flashcard Practice     30
Quiz     31
Test     32

**Moving On! Part 2 - Upper 6's, 7's, 8's & 9's**
Learn the Ten Stories     37-56
Crossword Puzzle Challenge     57
Story Review Quiz     59
Flashcard Practice     60
Quiz     61
Test     62

**Additional Reinforcements**
Part 1 Quiz & Test     64-65
Part 2 Quiz & Test     66-67
Master Test (Includes Parts 1 & 2 Combined)     68
Part 1 Cut-Out Flashcards     69-72
Part 2 Cut-Out Flashcards     73-76
Roll 'Em Cube Game Parts 1 & 2     77-79
Memory Story Booklet Parts 1 & 2     81-84

**Expert Level - Division**
Part 1 Division Cut-Out Flashcards     87-90
Part 1 Division Quiz & Test     91-92
Part 2 Division Cut-Out Flashcards     93-96
Part 2 Division Quiz & Test     97-98

**Answer Keys**     99-103

**BONUS DVD**
The bonus Times Tales® Game Show Quiz is a fun way to challenge the student on the stories and flashcards they learned in the workbook.

For more information and to see our Classroom Editions of Times Tales® please visit us at **www.TimesTales.com** or contact:
Trigger Memory Co. LLC
2395 Delle Celle Dr., Richland, WA 99354
Email: **TriggerMemory1@gmail.com**
Customer Service: 541-377-0064 Mon.-Fri. 9-5 (PST)

Trigger Memory Co. grants permission to photocopy ONLY the reproducible pages in the *back of this book* (pages 64-98). No other part of this publication may be reproduced in whole or part, transmitted or stored electronically or via any other means without the consent of the publisher.

WORKBOOK EDITION

# Introduction

Tired of boring flashcards that don't make the multiplication facts stick? Want a fun, unconventional way for your student to learn the upper times tables? Times Tales® is a creative, mnemonic (memory aid) system for students to memorize the most difficult math facts. With Times Tales® you can truly teach a student the upper times tables in a fraction of the time spent with traditional, rote memorization methods.

This system utilizes simple stories as triggers, or memory pegs, for a student to quickly recall the upper times tables. Times Tales® appeals to the visual, auditory, and kinesthetic learner. It has been proven successful for students with learning disabilities and those struggling to learn their times tables with traditional methods . . . and best of all, it's fun!

# Instructions

**HOW IT WORKS** - As the student works through the first part of the program, it is not necessary to know that the stories they are learning are really hidden multiplication problems. By the end of Part 1, everything will make sense, and the student will realize that the stories they have just learned can be converted into math facts. Not only is Times Tales® fun, but it will save hours of endless flashcard review. So ditch the drills, and let's get started!

**TWO PARTS** - In order to ensure the highest rate of success, the program is divided into two parts. Part 1 covers the upper 3's & 4's multiplication facts and includes eight stories. Part 2 covers the upper 6's, 7's, 8's, & 9's multiplication facts and contains ten stories. It is recommended for the student to wait one week after finishing Part 1 before "Moving On" to Part 2. During this time, the student can utilize the additional reinforcements (located in the back of the book) for review and mastery.

**ADDITIONAL REINFORCEMENTS** - Times Tales® has been successful with many types of learning styles. Some students may find they only need to work through Parts 1 & 2 of the main section of the workbook. However, other students may need a little more practice. The additional reinforcement section includes a fun game, a Memory Story Booklet, and a variety of flashcards and tests for review.

**EXPERT LEVEL (Division)** - Also included in the back of the workbook, is an "Expert Level" section. This section includes division to challenge students that have already mastered all the multiplication facts from Parts 1 & 2. The division tests and flashcards give the student a chance to try out their new skills with the division facts.

**BONUS TIMES TALES GAME SHOW DVD** - Times Tales® Workbook Edition comes with a bonus DVD to quiz the student on the Times Tales® stories and multiplication facts they have learned throughout the workbook. The DVD also includes a division portion of the game show quiz. This provides the student with an opportunity to test their skills on how quickly they can answer the multiplication and division problems as they are challenged to "Beat the Clock".

# A Peek Inside

WORKBOOK EDITION

**Learn the Story**
The first step is for the student to get to know the Times Tales® stories. Answers my vary during this step. One of the most effective ways for mnemonics to work, is to provide a "memory peg" to retain information. We have found that student retention of the Times Tales® stories increases when they express how each of the stories make them feel.

**Story Review**
The next page begins with the student writing the story in the <u>correct order</u> by unscrambling the words in the box. Once the story sentence is unscrambled, the student will then rewrite the sentence on the blank lines provided next to the Times Tales® story. Since the Times Tales® program is built upon knowing the stories, this step is very important and the sentences MUST be written in the proper order. This step also provides additional review of the key elements of each story.

**Additional Story Review**
To reinforce the stories there is a Crossword Puzzle and a Story Review Quiz at the end of each part.

**How it Works! Light Bulb Moment (page 29)**
This page explains to the student how the stories they just learned are really hidden multiplication problems. This is what we call the Times Tales® "Light Bulb Moment" when many students are amazed at how painless it was to learn their multiplication facts.

WORKBOOK EDITION

**Flashcard Quizzes & Tests**
Once the student knows how Times Tales® works, their new skill is challenged with a Quiz that uses the number symbol characters in place of the written numbers. They then can easily transition to using written numbers when taking the Test.

**Additional Reinforcements**
Some students may find they need additional review of the Times Tales® stories and multiplication facts. A variety of reinforcement reproducibles are available in the back of the workbook.

**Expert Level - Division**
This section is for students to apply their new multiplication skills from Parts 1 & 2, using division flashcards, quizzes, and tests.

**Answer Keys**
All the worksheets, quizzes, and tests have an answer key located in the back of the workbook. The beginning of Parts 1 & 2, "Learn the Story" & "Story Review" sections will need a parent's or teacher's review, as some of the answers may vary.

**Bonus DVD**
The Times Tales® Game Show Quiz DVD is a fun way to challenge students to apply their new multiplication skills while trying to "Beat the Clock". The DVD also includes a division section!

Part 1

# Let's Begin!

**IMPORTANT**: Do not proceed to Part 2 until ALL the multiplication facts have been mastered in Part 1. We recommend spending a few days utilizing the additional reinforcements (such as Bonus Game Show DVD, flashcards, tests, etc.) located in the back of the workbook BEFORE moving on to the new stories in Part 2.

Part 1

# Meet the Number Symbol Characters

The first step to learning the Times Tales® stories is to get to know each character and the number it represents.

 **Butterfly**

 **Chair**

 This boy represents the **Sixth Grade Class.**

 **Mrs. Week**
*(There is also a Mr. Week, but you will only see him if he is with his wife Mrs. Week.)*

 **Mrs. Snowman**
*(There is also a Mr. Snowman, but you won't see him unless he is with his wife, Mrs. Snowman.)*

 **Treehouse**

Part 1

# Number Symbol Review

The first step to learning the Times Tales® stories is to get to know each character and the number it represents.

1. Write the "hidden number" in the circle to the left of each number symbol.
2. Fill in the boxes with the name of each character.

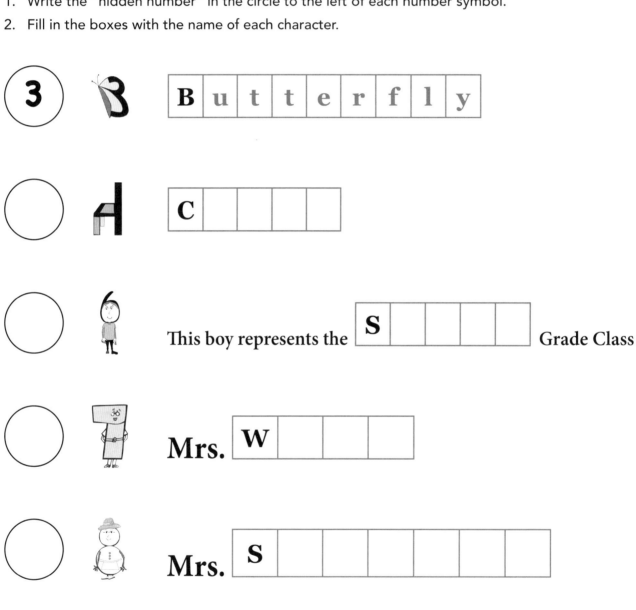

Part 1

# Step 1 — Learn the Story

Read each story, then answer the questions. Do not move on to Part 2 until ALL the stories have been learned and mastered. If more review is needed after finishing Part 1, there are additional reinforcements in the back of the workbook starting on page 64.

1. Read the story, and look at the picture.

The Sixth Grade Class played musical Chairs for 24 hours.

2. Twenty-four hours is an entire day. Explain how you would feel after playing musical Chairs for 24 hours.

_____
_____
_____
_____
_____

3. Why do you think the Sixth Grade Class decided to play musical Chairs for that many hours?

_____
_____
_____
_____
_____

Part 1

# Story Review

1. Using the words in the box, write a complete sentence making sure that the story is in the correct order.
   _____
   _____
   _____

   musical Chairs

   played

   for 24 hours.

   The Sixth Grade Class

2. Write the story sentence below.

   The _____
   _____
   _____
   _____
   _____
   _____

3. Choose the sentence below that is written correctly and in the correct order.
   ☐ The Fifth Grade Class played musical Chairs for 22 hours.
   ☐ The Sixth Grade Class played musical Chairs for 36 hours.
   ☐ The Sixth Grade Class played musical Chairs for 24 hours.

4. Who played musical Chairs? _____
   How long did they play?
   a. 36 hours    b. 24 hours    c. 12 hours

Part 1

# Learn the Story

1. Read the story, and look at the picture.

Mrs. Week went Butterfly hunting. She captured 20 in her net and 1 landed on her head.

2. Keeping 20 Butterflies in a net might be difficult. Explain how you would try to keep 20 Butterflies in a net and what you would do if one got loose.

_____

_____

_____

_____

3. Based upon the picture, do you think Mrs. Week knows that there is a Butterfly on her head? Why or why not?

_____

_____

_____

_____

4. With all the Butterflies in her net and the one that landed on her head, how many Butterflies does Mrs. Week have in all?_____

Part 1

1. Using the words in the box, write complete sentences making sure that the story is in the correct order.

   _____
   _____
   _____

   > Butterfly hunting.
   >
   > Mrs. Week went
   >
   > She captured 20 in her net
   >
   > and 1 landed on her head.

2. Write the story sentences below.

Mrs. _____
_____
_____
_____
_____
_____

3. Choose the line below that has the sentences written correctly and in the correct order.

   ☐ Mrs. Week went Butterfly hunting. She captured 21 in her net and 1 landed on her head.

   ☐ Mrs. Week went Butterfly hunting. She captured 20 in her net and 1 landed on her head.

   ☐ Mrs. Snowman went Butterfly hunting. They captured 20 in her net and 1 landed on her head.

Part 1

# Learn the Story

1. Read the story, and look at the picture.

Mrs. Snowman stood on a Chair to reach her 3 buttons and 2 mittens.

2. Explain why Mrs. Snowman might want to put on her buttons and mittens.

_____

_____

_____

_____

3. Based upon the picture, explain why Mrs. Snowman needs a Chair to get her buttons and mittens.

_____

_____

_____

_____

4. Fill in the blanks.

   Mrs. Snowman stood on a Chair to reach her____buttons and____mittens.

5. Draw a picture in each box to show the correct number of buttons and mittens Mrs. Week was trying to reach.

Buttons

Mittens

Part 1

# Story Review

1. Using the words in the box, write a complete sentence making sure that the story is in the correct order.

   _____

   _____

   _____

   Mrs. Snowman stood on a Chair

   and 2 mittens.

   3 buttons

   to reach her

2. Write the story sentence below.

Mrs._____

3. Choose the sentence below that is written correctly and in the correct order.

   ☐ Mrs. Snowman stood on a Chair to reach her 2 buttons and 3 mittens.

   ☐ Mrs. Week stood on a Chair to reach her 2 mittens and 3 buttons.

   ☐ Mrs. Snowman stood on a Chair to reach her 3 buttons and 2 mittens.

Part 1

# Learn the Story

1. Read the story, and look at the picture.

A Butterfly flew by the Treehouse and discovered a treasure of 27 cents.

2. Would you consider 27 cents a treasure? Why or why not?

   _____
   _____
   _____
   _____
   _____

3. Explain if you think the Butterfly will be able to carry the 27 cent treasure away.

   _____
   _____
   _____
   _____
   _____

Part 1

# Story Review

1. Using the words in the box, write a complete sentence making sure that the story is in the correct order.
   _____
   _____
   _____

   and discovered

   A Butterfly

   a treasure of 27 cents.

   flew by the Treehouse

2. Write the story sentence below.

A_____
_____
_____
_____
_____
_____

3. Choose the sentence below that is written correctly and in the correct order.

   ☐ A Butterfly flew by the Chair and discovered a treasure of 27 cents.

   ☐ A Butterfly flew by the Treehouse and discovered a treasure of 27 cents.

   ☐ Mrs. Snowman discovered 27 cents.

4. Fill in the blanks.
   One quarter and two pennies in the treasure chest is _____ cents.

Part 1

# Learn the Story

1. Read the story, and look at the picture.

Mrs. Week sat on a Chair and fished. She caught 2 boots and 8 trout.

2. Explain how you think those 2 boots ended up in the water.

   _____
   _____
   _____
   _____
   _____

3. Explain whether or not you would be able to eat all those trout for dinner.

   _____
   _____
   _____
   _____

4. Draw a picture in each box to show the correct number of boots and trout that Mrs. Week caught.

Boots

Trout

Part 1

 Story Review

1. Using the words in the box, write complete sentences making sure that the story is in the correct order.

   _____
   _____
   _____

   > She caught 2 boots
   > a Chair and fished.
   > Mrs. Week sat on
   > and 8 trout.

2. Write the story sentences below.

Mrs. _____
_____
_____
_____
_____

3. Choose the line below that has the sentences written correctly and in the correct order.

   ☐ Mrs. Week sat on a Chair and fished. She caught 2 boots and 8 trout.

   ☐ Mrs. Week sat on a Chair and fished. She caught 3 boots and 9 trout.

   ☐ Mrs. Week sat on a Chair and fished. She caught 4 boots and 8 trout.

Part 1

# Learn the Story

1. Read the story, and look at the picture.

The Sixth Grade Class raised Butterflies. At 1 o'clock, they set 8 Butterflies free.

2. Why do you think the Sixth Grade Class decided to let the 8 Butterflies free at 1 o'clock?

_____
_____
_____
_____
_____
_____

3. Do you think the 8 Butterflies will stay together once they fly out the window? Why or why not?

_____
_____
_____
_____

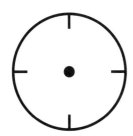

Draw the time on the clock the Butterflies were set free.

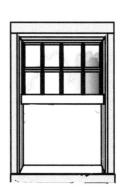

Draw a group of 8 Butterflies heading towards the open window.

Part 1

  Story Review

1. Using the words in the box, write complete sentences making sure that the story is in the correct order.

   _____

   _____

   _____

   > set 8 Butterflies free.
   >
   > raised Butterflies.
   >
   > The Sixth Grade Class
   >
   > At 1 o'clock, they

2. Write the story sentences below.

The _____

_____

_____

_____

_____

3. Choose the line below that has the sentences written correctly and in the correct order.

   ☐ The Fifth Grade Class raised Butterflies. At 1 o'clock, they set 6 Butterflies free.

   ☐ The Sixth Grade Class raised Butterflies. At 2 o'clock, they set 8 Butterflies free.

   ☐ The Sixth Grade Class raised Butterflies. At 1 o'clock, they set 8 Butterflies free.

4. Fill in the blanks.

   What time were the Butterflies set free? ___ o'clock

   How many Butterflies were set free? ____

Part 1

# Learn the Story

1. Read the story, and look at the picture.

The Chair got stuck in the Treehouse after someone tied on 3 kites and 6 balloons.

2. Explain why you think someone might tie the 3 kites and 6 balloons onto a Chair.

   _____
   _____
   _____
   _____
   _____

3. Would you climb up to the Treehouse to get the Chair untangled? Explain your answer.

   _____
   _____
   _____
   _____

4. Draw a picture in each box to show the correct number of kites and balloons that were tied to the Chair.

Kites

Balloons

Part 1

# Story Review

1. Using the words in the box, write a complete sentence making sure that the story is in the correct order.

   _____
   _____
   _____

   > The Chair got stuck
   >
   > in the Treehouse
   >
   > 3 kites and 6 balloons.
   >
   > after someone tied on

2. Write the story sentence below.

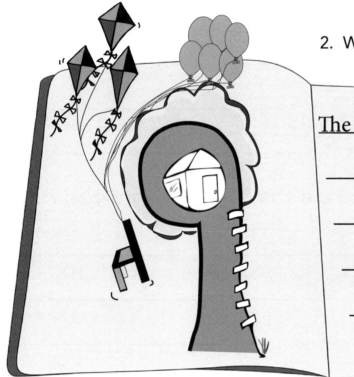

   The _____
   _____
   _____
   _____
   _____

3. Choose the sentence below that is written correctly and in the correct order.

   ☐ The Chair got stuck in the Treehouse after someone tied on 3 kites and 4 balloons.

   ☐ The Chair got stuck in the Treehouse after someone tied on 2 kites and 3 balloons.

   ☐ The Chair got stuck in the Treehouse after someone tied on 3 kites and 6 balloons.

Part 1

# Learn the Story

1. Read the story, and look at the picture.

After the warm weather melted Mrs. Snowman, 2 Butterflies took her arms and 4 flew away with her apron.

2. Warm weather brings out the Butterflies, but it also melts the snow. Explain what you think the Butterflies plan on doing with the stick arms and apron.

_____

_____

_____

_____

3. Explain if you think 4 Butterflies would be able to fly away with an apron.

_____

_____

_____

_____

4. Draw in the correct number of Butterflies on each picture below.

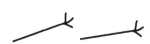   

25

Part 1

 Story Review

1. Using the words in the box, write a complete sentence making sure that the story is in the correct order.

   _____
   _____
   _____

weather melted Mrs. Snowman,

2 Butterflies took her arms

After the warm

and 4 flew away with her apron.

2. Write the story sentence below.

After _____
_____
_____
_____
_____
_____

3. Choose the sentence below that is written correctly and in the correct order.

   ☐ After the warm weather melted Mrs. Snowman, 2 Butterflies took her apron and 3 flew away with her arms.

   ☐ After the warm weather melted Mrs. Snowman, 2 Butterflies took her arms and 4 flew away with her apron.

   ☐ After the warm weather melted Mrs. Snowman, 3 Butterflies took her arms and 4 flew away with her apron.

Part 1

# CROSSWORD PUZZLE CHALLENGE

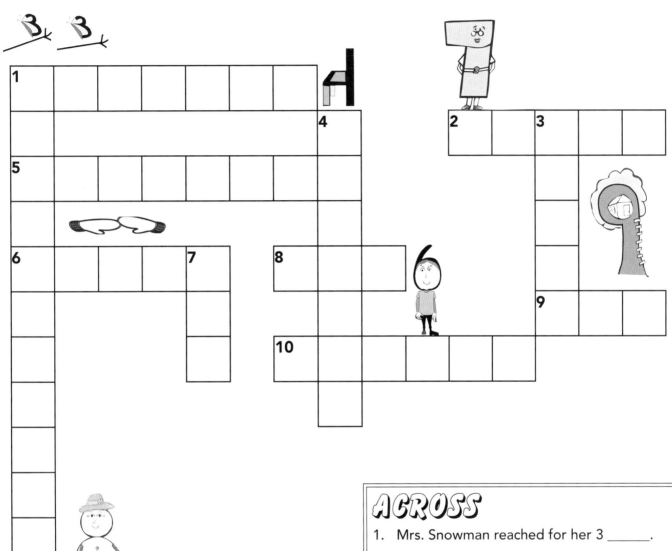

## ACROSS

1. Mrs. Snowman reached for her 3 _____.
2. The Sixth Grade _____.
5. The Butterfly discovered a _____ at the Treehouse.
6. How many Butterflies were set free?
8. The Butterflies were set free at __ o'clock.
9. Mrs. Week captured Butterflies in her _____.
10. The warm weather _____ Mrs. Snowman.

## DOWN

1. What was in the net?
3. ...and 4 flew away with her _____.
4. The treasure was 1 quarter and 2 _____.
7. How many boots did Mrs. Week catch?

Part 1

# Story Review Quiz

Write the correct answer in the blank.

1. How many Butterflies flew away with her arms? ____
   How many flew away with the apron? ____

2. How many hours did the Sixth Grade Class play musical Chairs? ____

3. Who found the treasure? _____
   Where was it found? _____

4. What time were the Butterflies set free? ___ o'clock
   How many were set free? ____

5. Who stood on a Chair to reach something? Mrs. _____

6. How many boots did Mrs. Week catch? ____
   How many trout did she catch?____

7. The Chair was attached to how many kites? ____
   How many balloons were attached? ____

8. How many Butterflies did Mrs. Week have in her net? ____
   How many did she have on her head? ____

9. How many buttons were on the shelf? ____
   How many mittens were on the shelf? ____

10. Who melted when the warm weather came? Mrs._____

11. How much money was in the treasure chest? ____cents

12. Who played musical Chairs? _____

# LIGHT BULB MOMENT!

### See How it Works

1. Did you know there is a multiplication problem hiding in each Times Tales® story you just learned? Now that you have learned the stories, you will be able to answer the hidden multiplication problems.

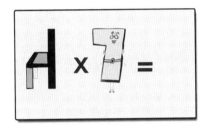

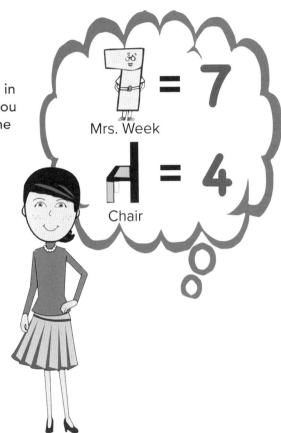

**This flashcard is for the multiplication problem: 4x7 or 7x4**

2. To solve this multiplication problem all you have to do is to remember the Times Tales® story that has Mrs. Week and the Chair. There was only one story that had both of those number symbols.

 x  =

**Mrs. Week** sat on a **Chair** and fished. She caught **2** boots and **8** trout.

### 7 x 4 = 2 8

Part 1

Step 2 Flashcard Practice

Name:_____
Time:_____
Score:_____

Draw a line to connect the correct missing parts in each Times Tales® story. Then write the correct answers in the shaded boxes.

| | | |
|---|---|---|
| 7 × 4 = | 🎏 (fishing) | 2  8 |
| 8 × 9 = | | |
| 6 × 4 = | | |
| 7 × 8 = | | |
| 6 × 4 = | | |
| 9 × 4 = | | |
| 6 × 8 = | | |
| 6 × 8 = | | |

Part 1

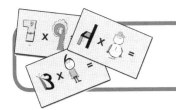

# Quiz

Name:_____
Time:_____
Score:_____

Write the correct answer for each practice problem in the shaded boxes.
*Hint*: Remember the story that has both of the number symbols pictured.

Part 1

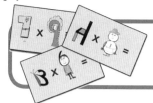

# Test

Name:_____
Time:_____
Score:_____

Write the correct answer on each line.

*Tip!* If you miss any, you will need to go back and review the stories by utilizing the DVD and additional worksheets (starting on page 64). *Do not move onto Part 2 until ALL the multiplication problems have been mastered.

| 7 x 4 = | 4 x 8 = |
| 3 x 9 = | 3 x 6 = |
| 9 x 4 = | 4 x 6 = |
| 6 x 3 = | 3 x 8 = |
| 8 x 3 = | 4 x 9 = |
| 6 x 4 = | 9 x 3 = |
| 7 x 3 = | 4 x 7 = |
| 8 x 4 = | 3 x 7 = |

 Challenge your skills further with the Times Tales® Game Show Quiz DVD

Part 1

# Additional Reinforcements

**\*\*Do not move on to Part 2, until ALL of the multiplication facts have been mastered.** We recommend spending one week utilizing the additional reinforcements of Part 1 (*Game Show Quiz DVD, worksheets, Cube Game, etc.*) **before** beginning Part 2.

## Part 1 Review

| | |
|---|---|
| Additional Quiz & Test | 64-65 |
| Cut-Out Flashcards | 69-72 |
| Roll 'Em Cube Game | 77 |
| Memory Story Booklet | 81-82 |

**Expert Level**

| | |
|---|---|
| Division Cut-Out Flashcards | 87-90 |
| Division Quiz & Test | 91-92 |

**DVD Game Show Quiz DVD**

*Trigger Memory Co.© grants permission to photocopy reproducibles located in the back of this workbook (pages 64-98) for use within one household or classroom. No other part or whole of the Times Tales® Workbook may be copied without the written consent of the publishers.*

Part 1

# Draw

**Draw one of the stories from Part 1.**

Write the story sentence for the picture.

_____

_____

Write the multiplication problem and answer for the story you drew.

 X  =

Part 2

# Moving On!

8x6   6x6
7x7   7x9   9x6
8x9
6x7   7x8
8x8   9x9

**IMPORTANT** Do not proceed to Part 2 until ALL the multiplication facts have been mastered in Part 1. We recommend spending a few days utilizing the additional reinforcements (such as Bonus Game Show DVD, flashcards, tests, etc.) located in the back of the workbook BEFORE moving on to the new stories in Part 2.

Part 2

# Learn the Story

1. Read the story, and look at the picture.

Mr. and Mrs. Week have 4 dogs and 9 cats.

2. Do you think those 4 dogs and 9 cats get along? Explain your answer.

_____
_____
_____
_____
_____

3. Maybe Mr. and Mrs. Week have only 4 dogs, compared to 9 cats, because dogs are much harder to take care of than cats. Do you agree? Explain your answer.

_____
_____
_____
_____

4. Draw 4 dogs in the first box and 9 cats in the second one.

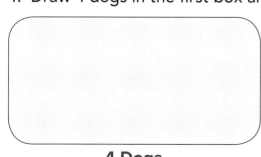

**4 Dogs**

**9 Cats**

Part 2

Story Review

1. Using the words in the box, write a complete sentence making sure that the story is in the correct order.

   _____
   _____
   _____

   and 9 cats.

   Mr. and Mrs. Week

   have 4 dogs

2. Write the story sentence below.

Mr. _____

3. Choose the sentence below that is written correctly and in the correct order.

   ☐ Mr. and Mrs. Week have 6 dogs and 8 cats.

   ☐ Mr. and Mrs. Week have 4 dogs and 9 cats.

   ☐ Mr. and Mrs. Week have 4 dogs and 6 cats.

Part 2

# Learn the Story

1. Read the story, and look at the picture.

Mrs. Week and Mrs. Snowman were driving together and went one mile over the speed limit.

2. Do you think Mrs. Week and Mrs. Snowman might get a speeding ticket for going only one mile over the speed limit? Why or why not?

_____

_____

_____

_____

_____

3. Why do you think Mrs. Snowman and Mrs. Week didn't notice they were going one mile over the speed limit? Explain your answer.

_____

_____

_____

4. Fill in the sign and speedometer below.

Write the speed limit in the sign.

Write in the box how fast Mrs. Week and Mrs. Snowman were driving.

39

Part 2

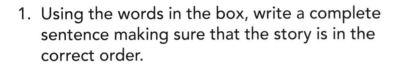

1. Using the words in the box, write a complete sentence making sure that the story is in the correct order.

   _____

   _____

   _____

   > and Mrs. Snowman
   >
   > and went one mile over the speed limit.
   >
   > were driving together
   >
   > Mrs. Week

2. Write the story sentence below.

   Mrs. _____

   _____

   _____

   _____

   _____

   _____

3. Choose the sentence below that is written correctly and in the correct order.

   ☐ Mrs. Week and Mrs. Snowman were driving together and went 10 miles over the speed limit.

   ☐ Mrs. and Mr. Week were driving together and went one mile over the speed limit.

   ☐ Mrs. Week and Mrs. Snowman were driving together and went one mile over the speed limit.

4. How fast were Mrs. Week and Mrs. Snowman going?
   ☐ 55
   ☐ 56
   ☐ 57

Part 2

# Learn the Story

1. Read the story, and look at the picture.

Mrs. Week's Sixth Grade Class has 42 students.

2. A classroom of 42 students might cause some problems for Mrs. Week. List two problems Mrs. Week might have teaching a class that large.

   1. _____
   _____

   2. _____
   _____

3. Would you like to be in a classroom of 42 students? Explain your answer.

_____
_____
_____
_____

Part 2

 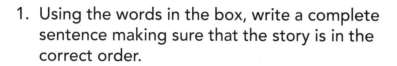

1. Using the words in the box, write a complete sentence making sure that the story is in the correct order.

   _____

   _____

   _____

   has 42 students.

   Sixth Grade Class

   Mrs. Week's

2. Write the story sentence below.

Mrs. _____

3. Choose the sentence below that is written correctly and in the correct order.
   ☐ Mrs. Week's Sixth Grade Class has 40 students.
   ☐ Mrs. Week's Sixth Grade Class has 41 students.
   ☐ Mrs. Week's Sixth Grade Class has 42 students.

4. Fill in the blanks
   Mrs. _____ _____ Grade Class has _____ students.

Part 2

# Learn the Story

1. Read the story, and look at the picture.

Two Sixth Grade Classes played soccer and the score was 3 to 6.

2. Based upon the picture, do you think the Sixth Grade Classes are playing during recess just for fun, or are the teams from two different schools?  Explain your answer.

_____

_____

_____

_____

3. Do you think the team that only has 3 points has a chance to win the game? Explain your answer.

_____

_____

_____

_____

4. Fill in the scoreboard below.

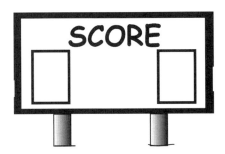

43

Part 2

 Story Review

1. Using the words in the box, write a complete sentence making sure that the story is in the correct order.

   _____
   _____
   _____

   > Two Sixth Grade Classes
   >
   > was 3 to 6.
   >
   > played soccer
   >
   > and the score

2. Write the story sentence below.

   Two_____
   _____
   _____
   _____
   _____
   _____

3. Choose the sentence below that is written correctly and in the correct order.
   ☐ Two Sixth Grade Classes played soccer and the score was 3 to 6.
   ☐ Two Sixth Grade Classes played soccer and the score was 2 to 6.
   ☐ Two Sixth Grade Classes played soccer and the score was 6 to 6.

Part 2

# Learn the Story

1. Read the story, and look at the picture.

Mrs. Week went to the Treehouse and raked up 6 bags of leaves by 3 o'clock.

2. Do you think raking 6 bags of leaves is a big job or a task that can be completed quickly? Explain your answer.

_____
_____
_____
_____
_____

3. Mrs. Week finished raking the leaves at 3 o' clock. Explain what you do every day at 3 o'clock.

_____
_____
_____
_____

4. Draw the correct number of bags of leaves that Mrs. Week raked and write the time on the clock when she finished.

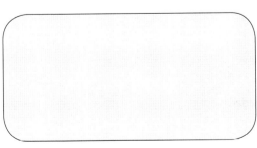

**Bags of leaves**

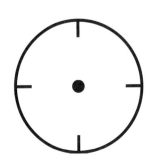

45

Part 2

 Story Review

1. Using the words in the box, write a complete sentence making sure that the story is in the correct order.

   _____
   _____
   _____

   6 bags of leaves

   Mrs. Week went to the Treehouse

   and raked up

   by 3 o'clock.

   2. Write the story sentence below.

Mrs. _____

3. Choose the sentence below that is written correctly and in the correct order.
   ☐ Mrs. Week went to the Treehouse and raked up 4 bags of leaves by 3 o'clock.
   ☐ Mr. Week raked up 6 bags of leaves by 3 o'clock.
   ☐ Mrs. Week went to the Treehouse and raked up 6 bags of leaves by 3 o'clock.

4. Fill in the blanks.
   How many bags of leaves did Mrs. Week rake? ___
   What time did she finish raking? ___ o' clock

Part 2

## Learn the Story

1. Read the story, and look at the picture.

Mr. and Mrs. Snowman must eat 6 snow cones 4 times a day.

2. In order for Mr. and Mrs. Snowman to avoid melting, they have to keep eating snow cones throughout the day. How does your body feel when you eat something very cold? Explain your answer.

_____

_____

_____

3. Mr. and Mrs. Snowman had to eat the 6 snow cones 4 times a day. That would be breakfast, lunch, dinner, and a snack. Would you still be hungry if you only ate snow cones all day long? Explain your answer.

_____

_____

_____

4. Draw the number of snow cones Mr. and Mrs. Snowman must eat each meal, then answer the question.

How many times a day did they have to eat the snow cones? _____

☑ Breakfast
☑ Lunch
☑ Dinner
☑ Snack

Draw the snow cones

Part 2

# Story Review

1. Using the words in the box, write a complete sentence making sure that the story is in the correct order.

   _____
   _____
   _____

   > Mr. and Mrs. Snowman
   >
   > 4 times a day.
   >
   > must eat
   >
   > 6 snow cones

2. Write the story sentence below.

Mr._____
_____
_____
_____
_____
_____

3. Choose the sentence below that is written correctly and in the correct order.
   - ☐ Mr and Mrs. Week must eat 6 snow cones 4 times a day.
   - ☐ Mr. and Mrs. Snowman must eat 6 snow cones 4 times a day.
   - ☐ Mr. and Mrs. Snowman must eat 4 snow cones 6 times a day.

4. Fill in the blanks.

   How many snow cones did Mr. and Mrs. Snowman have to eat each meal? ____
   How many times a day did they have to eat the snow cones? ____

Part 2

# Learn the Story

1. Read the story, and look at the picture.

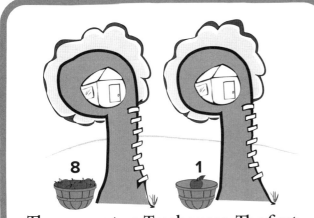

There were two Treehouses. The first Treehouse grew 8 apples and the second grew only 1 apple.

2. What do you think caused the first Treehouse to grow 8 apples but the second to grow only one apple? Explain your answer.

_____

_____

_____

_____

_____

3. Do you think having a Treehouse in an apple tree would be a problem? Explain your answer.

_____

_____

_____

_____

4. Draw the correct number of apples in the baskets below.

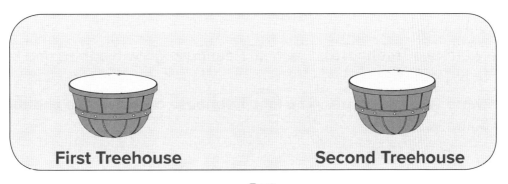

First Treehouse          Second Treehouse

Part 2

# Story Review

1. Using the words in the box, write complete sentences making sure that the story is in the correct order.

   _____

   _____

   _____

   > There were two Treehouses.
   >
   > and the second
   >
   > grew only 1 apple.
   >
   > The first Treehouse grew 8 apples

2. Write the story sentences below.

There _____

_____

_____

_____

_____

3. Choose the line below that has the sentences written correctly and in the correct order.

   ☐ There were two Treehouses. The first Treehouse grew 6 apples and the second grew only 1 apple.

   ☐ There were two Treehouses. The first Treehouse grew 8 apples and the second grew only 1 apple.

   ☐ There were two Treehouses. The first Treehouse grew 1 apple and the second grew 8 apples.

Part 2

# Learn the Story

1. Read the story, and look at the picture.

Mrs. Snowman visited the Sixth Grade Class and showed them 4 birds and an 8 legged spider.

2. Do you think those 4 birds and the 8 legged spider will stay on Mrs. Snowman's arms? Explain your answer.

_____

_____

_____

_____

_____

3. Would you like to have 4 birds and an 8 legged spider on your arms? Explain how that would make you feel.

_____

_____

_____

4. Draw the birds and spider on Mrs. Snowman's arms.

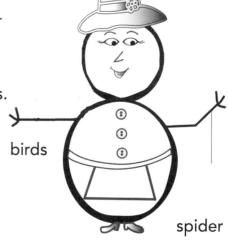

birds

spider

51

Part 2

 # Story Review

1. Using the words in the box, write a complete sentence making sure that the story is in the correct order.

   _____

   _____

   _____

   > and showed them 4 birds
   >
   > the Sixth Grade Class
   >
   > and an 8 legged spider.
   >
   > Mrs. Snowman visited

2. Write the story sentence below.

Mrs. _____

_____

_____

_____

_____

_____

3. Choose the sentence below that is written correctly and in the correct order.

   ☐ Mrs. Snowman visited the Sixth Grade Class and showed them an 8 legged spider and 4 birds.

   ☐ Mrs. Snowman visited the Sixth Grade Class and showed them 4 birds and an 8 legged spider.

   ☐ Mrs. Week visited the Sixth Grade Class and showed them 4 birds and an 8 legged spider.

Part 2

# Learn the Story

1. Read the story, and look at the picture.

Mrs. Snowman went to the Treehouse and knocked 7 times on 2 doors.

2. Maybe Mrs. Snowman is knocking on the door because she is visiting a friend. Why do you think she is knocking on the door? Explain your answer.

_____

_____

_____

_____

3. Do you know why she knocked 7 times on 2 doors? There is also a back door. If you knocked as many as 7 times on the front door and no one answered, would you then knock that many times on the back door as well? Explain your answer.

_____

_____

_____

_____

4. Fill in the blanks.

How many times did Mrs. Snowman knock on the doors? ____

How many doors did Mrs. Snowman knock on? ____

Part 2

# Story Review

1. Using the words in the box, write a complete sentence making sure that the story is in the correct order.

   _____
   _____
   _____

   Mrs. Snowman

   and knocked 7 times

   on 2 doors.

   went to the Treehouse

2. Write the story sentence below.

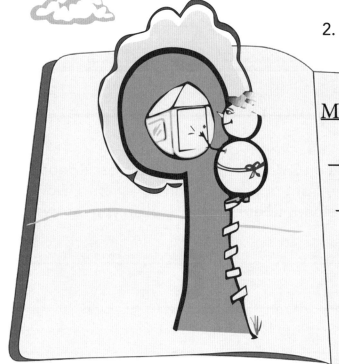

Mrs. _____
_____
_____
_____
_____
_____

3. Choose the sentence below that is written correctly and in the correct order.
   ☐ Mrs. Week went to the Treehouse and knocked 7 times on 2 doors.
   ☐ Mrs. Snowman went to the Treehouse and knocked 2 times on 7 doors.
   ☐ Mrs. Snowman went to the Treehouse and knocked 7 times on 2 doors.

Part 2

# Learn the Story

1. Read the story, and look at the picture.

The Sixth Grade Class went to the Treehouse to feed 5 pounds of bananas to 4 monkeys.

2. Do you think 5 pounds of bananas will be enough to feed 4 monkeys? Explain your answer.

   _____
   _____
   _____
   _____
   _____

3. Based upon the picture, do you think the Sixth Grade Class has brought bananas to the monkeys before? Why or why not?

   _____
   _____
   _____
   _____

4. Fill in the blanks.

   How many pounds of bananas did the Sixth Grade Class bring? _____
   How many monkeys were at the Treehouse? _____

Part 2

# Story Review

1. Using the words in the box, write a complete sentence making sure that the story is in the correct order.

   _____
   _____
   _____

   > The Sixth Grade Class
   >
   > 5 pounds of bananas
   >
   > went to the Treehouse to feed
   >
   > to 4 monkeys.

2. Write the story sentence below.

The _____

3. Choose the sentence below that is written correctly and in the correct order.

   ☐ The Sixth Grade Class went to the Treehouse to feed 5 pounds of bananas to 4 monkeys.

   ☐ The Sixth Grade Class went to the Treehouse to feed 4 pounds of bananas to 5 monkeys.

   ☐ The Sixth Grade Class went to the Treehouse to feed 4 pounds of bananas to 4 monkeys.

# CROSSWORD PUZZLE CHALLENGE

Part 2

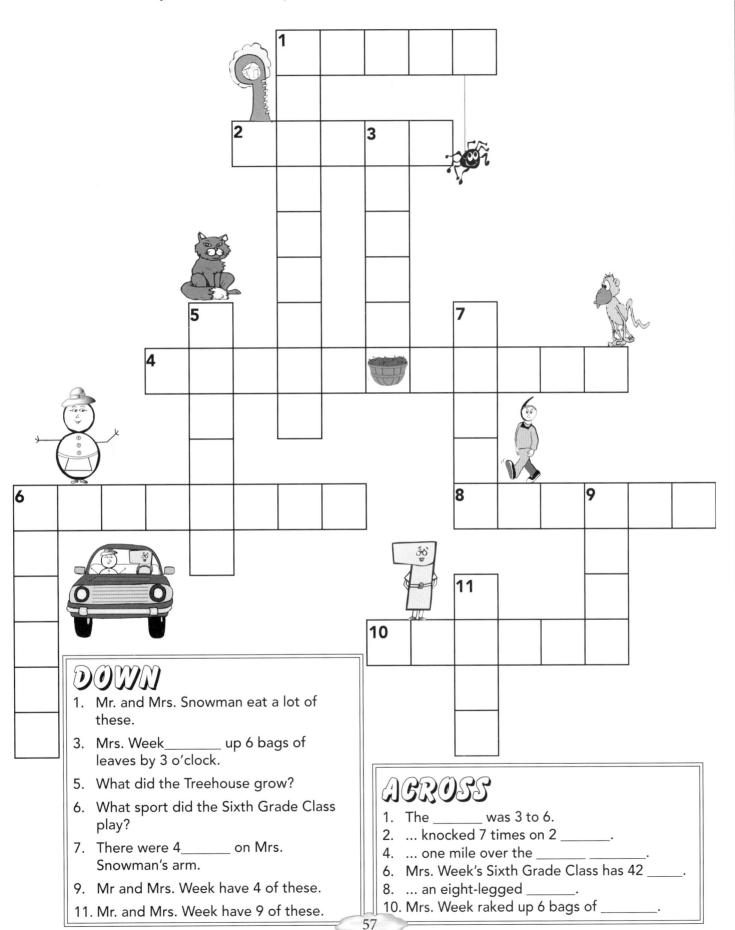

## DOWN

1. Mr. and Mrs. Snowman eat a lot of these.
3. Mrs. Week_____ up 6 bags of leaves by 3 o'clock.
5. What did the Treehouse grow?
6. What sport did the Sixth Grade Class play?
7. There were 4_____ on Mrs. Snowman's arm.
9. Mr and Mrs. Week have 4 of these.
11. Mr. and Mrs. Week have 9 of these.

## ACROSS

1. The _____ was 3 to 6.
2. ... knocked 7 times on 2 _____.
4. ... one mile over the _____ _____.
6. Mrs. Week's Sixth Grade Class has 42 _____.
8. ... an eight-legged _____.
10. Mrs. Week raked up 6 bags of _____.

Part 2

# Draw

**Draw one of the stories from Part 2.**

Write the story sentence for the picture.

_____

_____

Write the multiplication problem and answer for the story you drew.

 x  =

Part 2

# Story Review Quiz

Write the correct answer in the blank.

1. Who ate the snow cones? Mr. & Mrs. _____

2. Who went 56 miles an hour? Mrs._____ & Mrs. _____

3. Who fed the monkeys? _____

4. Where did Mrs. Week go to rake the leaves? _____

5. What grew apples? two _____

6. What did Mrs. Snowman show the Sixth Grade Class? _____ birds and an ___ legged spider

7. Mrs. Week raked how many bags of leaves? ____ What time did she finish? ___o'clock

8. How many snow cones did Mr. and Mrs. Snowman eat each meal? ___
   How many times a day did they eat them? _____

9. How many dogs do Mr. & Mrs. Week have? ____How many cats? ____

10. What was the score of the soccer game? ____ to ____

11. How many students are in Mrs. Week's class? _____

12. Where did Mrs. Snowman knock on the doors? _____
    How many times did she knock? ____
    On how many doors did she knock on? ____

13. How many pounds of bananas did the Sixth Grade Class bring?_____
    How many monkeys were at the Treehouse ready to eat bananas? ____

14. How many apples did the first Treehouse grow?_____
    How many apples did the second Treehouse grow? _____

15. Who went to the Sixth Grade Class to show them her pets?
    Mrs._____

Part 2

# Step 2 Flashcard Practice

Name:____________________
Time:_____
Score:_____

Draw a line to connect the missing parts in each Times Tales® story. Then write the answers in the shaded boxes.

☑ Breakfast
☑ Lunch
☑ Dinner
☑ Snack

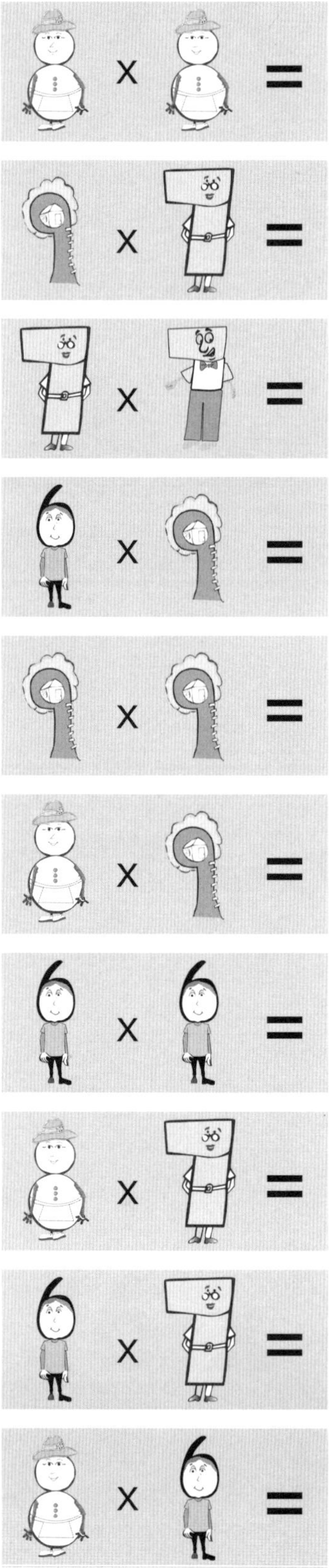

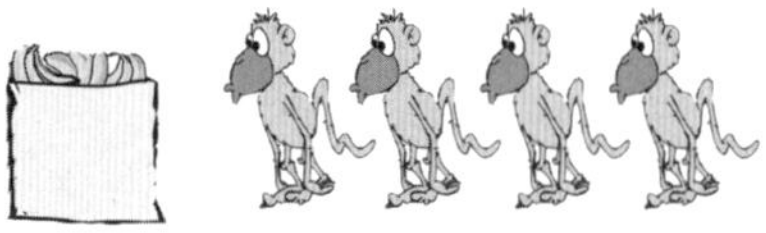

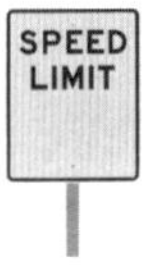

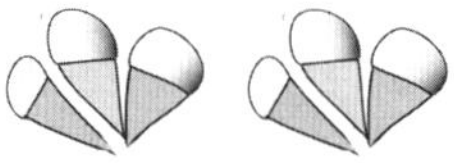

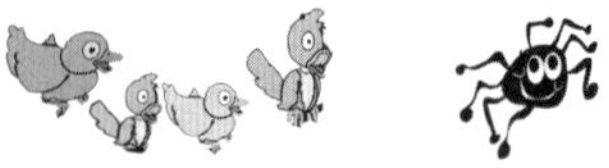

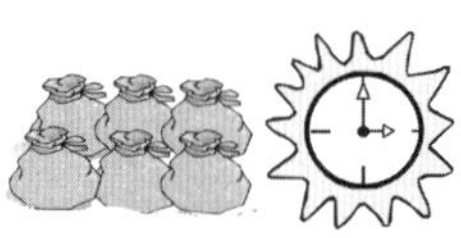

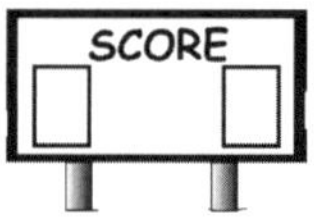

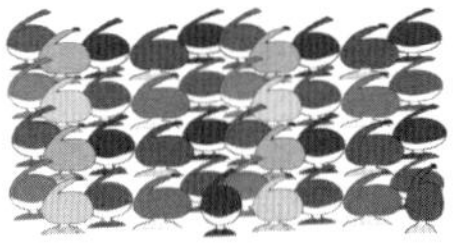

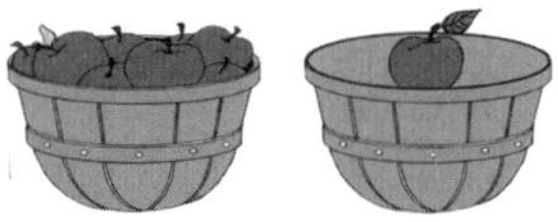

8 1

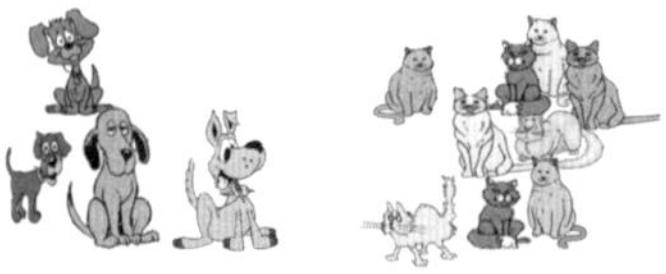

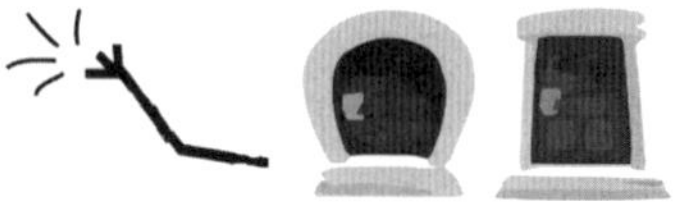

Part 2

# Quiz

Name:_____
Time:____
Score:____

Write the correct answer for each practice problem in the shaded boxes.
*Hint*: Remember the story that has both of the number symbols pictured.

**Part 2**

**Test**

Name:_____
Time:_____
Score:_____

Write the correct answer on each line.
*Tip!* If you miss any, you will need to go back and review the stories by utilizing the DVD and additional worksheets (starting on page 66).

6 x 6 =

7 x 9 =

6 x 7 =

8 x 9 =

7 x 7 =

9 x 9 =

6 x 9 =

7 x 8 =

9 x 8 =

8 x 6 =

9 x 7 =

9 x 6 =

8 x 7 =

8 x 8 =

7 x 6 =

6 x 8 =

Challenge your skills further with the Times Tales® Game Show Quiz DVD

Part 2

# Additional Reinforcements

You will find a Master Test with ALL the Times Tales® from Part 1 & 2 on page 68. You also might want to test your new multiplication skills with our Expert Division Level starting on page 85.

## Part 2 Review

| | |
|---|---|
| Additional Quiz & Test | 66-67 |
| Cut-Out Flashcards | 73-76 |
| Roll 'Em Cube Game | 79 |
| Memory Story Booklet | 83-84 |

**Expert Level**

| | |
|---|---|
| Division Cut-Out Flashcards | 93-96 |
| Division Quiz & Test | 97-98 |

**DVD Game Show Quiz DVD**

*Trigger Memory Co.© grants permission to photocopy reproducibles located in the back of this workbook (pages 64-98) for use within one household or classroom. No other part or whole of the Times Tales® Workbook may be copied without the written consent of the publishers.*

# Part 1

## Additional Quiz Part 1

Name:_____
Time:_____
Score:_____

Write the correct answer for each practice problem in the shaded boxes.
*Hint*: Remember the story that has both of the number symbols pictured.

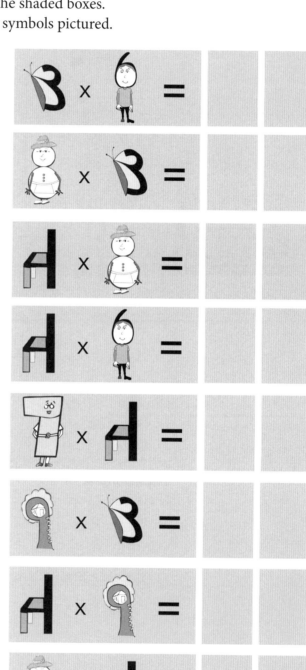

# Additional Test Part 1

Part 1

Name:_____
Time:_____
Score:____

Write the correct answer on each line.

3 x 7 =

4 x 9 =

9 x 3 =

4 x 7 =

8 x 4 =

6 x 4 =

3 x 6 =

8 x 3 =

6 x 3 =

9 x 4 =

4 x 6 =

3 x 8 =

3 x 9 =

7 x 3 =

7 x 4 =

4 x 8 =

Part 2

# Additional Quiz
# Part 2

Name:_____
Time:_____
Score:_____

Write the correct answer for each practice problem in the shaded boxes.
*Hint*: Remember the story that has both of the number symbols pictured.

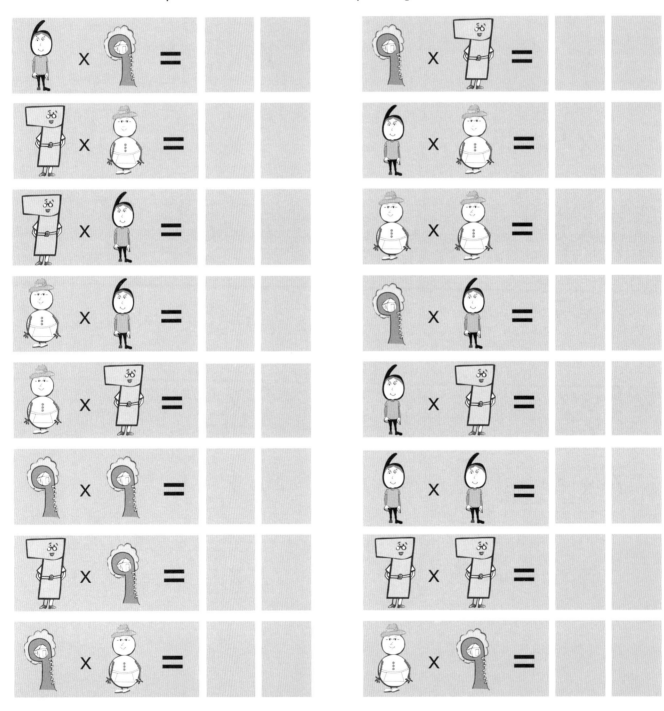

# Additional Test Part 2

Part 2

Name:_____
Time:_____
Score:____

Write the correct answer on each line.

7 x 6 =              6 x 7 =

6 x 9 =              7 x 7 =

9 x 8 =              9 x 7 =

8 x 6 =              7 x 8 =

6 x 6 =              8 x 7 =

8 x 8 =              9 x 9 =

8 x 9 =              6 x 8 =

9 x 6 =              7 x 9 =

# Master Test Part 1 & Part 2

Name:_____
Time:_____
Score:_____

4 x 6 =

6 x 6 =

7 x 9 =

6 x 7 =

3 x 9 =

8 x 9 =

7 x 4 =

6 x 8 =

3 x 6 =

8 x 4 =

7 x 3 =

4 x 9 =

7 x 7 =

9 x 6 =

8 x 7 =

8 x 8 =

8 x 3 =

9 x 9 =

Now test your skills with Division! Page 85

Part 1

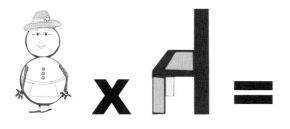

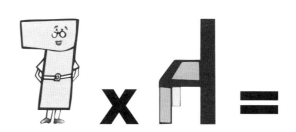

Part 1 — Cut Along Dotted Lines

| $4 \times 9 =$ | $3 \times 6 =$ |
| $7 \times 3 =$ | $8 \times 4 =$ |
| $9 \times 3 =$ | $4 \times 6 =$ |
| $7 \times 4 =$ | $6 \times 3 =$ |

Part 1

✂ Cut Along Dotted Lines

 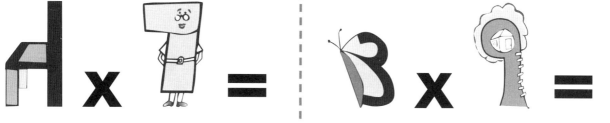

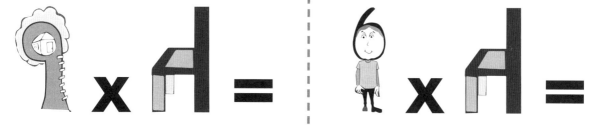 

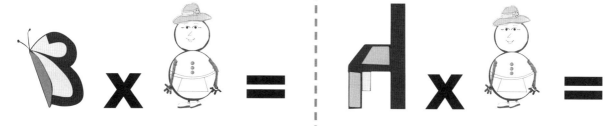 

71

Part 1  ✂ Cut Along Dotted Lines

| 8 x 3 = | 3 x 7 = |
| --- | --- |
| 3 x 9 = | 4 x 7 = |
| 6 x 4 = | 9 x 4 = |
| 4 x 8 = | 3 x 8 = |

Part 2

 Cut Along Dotted Lines

Part 2  ✂ Cut Along Dotted Lines

| 8 x 7 = | 7 x 8 = |
| 8 x 8 = | 8 x 6 = |
| 7 x 7 = | 6 x 7 = |
| 6 x 6 = | 7 x 6 = |

Part 2

✂ Cut Along Dotted Lines

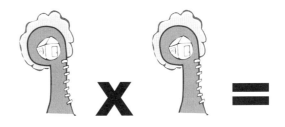

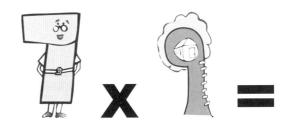

75

Part 2

| 9 x 7 = | 9 x 6 = |
| --- | --- |
| 9 x 9 = | 9 x 8 = |
| 7 x 9 = | 6 x 9 = |
| 6 x 8 = | 8 x 9 = |

Part 1

# ROLL-'EM CUBE GAME

## HOW TO MAKE PAPER CUBES:

1. Cut along edges of cube patterns.
2. Fold along black lines and form cubes.
3. Tape or glue flaps into place

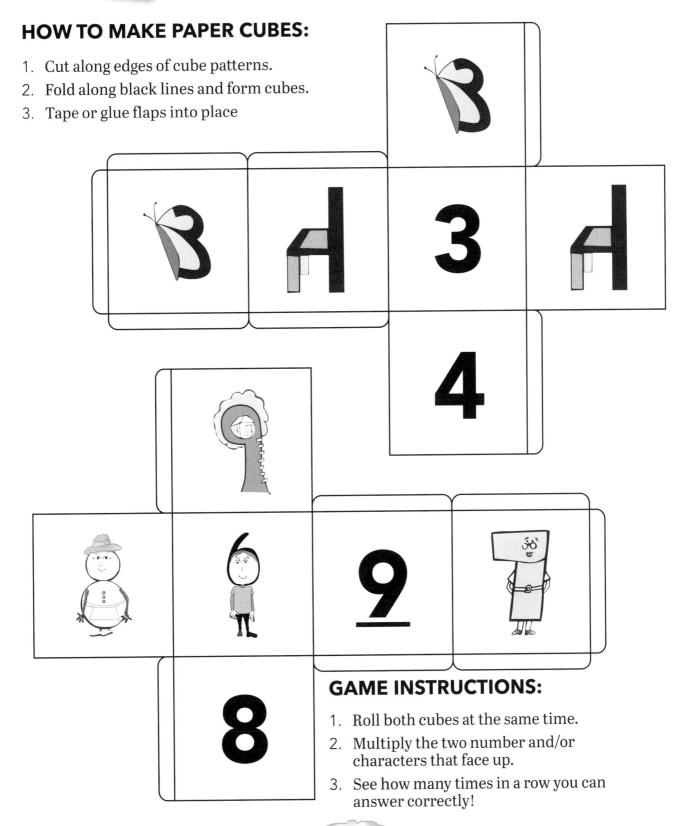

## GAME INSTRUCTIONS:

1. Roll both cubes at the same time.
2. Multiply the two number and/or characters that face up.
3. See how many times in a row you can answer correctly!

77

Part 2

## HOW TO MAKE PAPER CUBES:

1. Cut along edges of cube patterns.
2. Fold along black lines and form cubes.
3. Tape or glue flaps into place

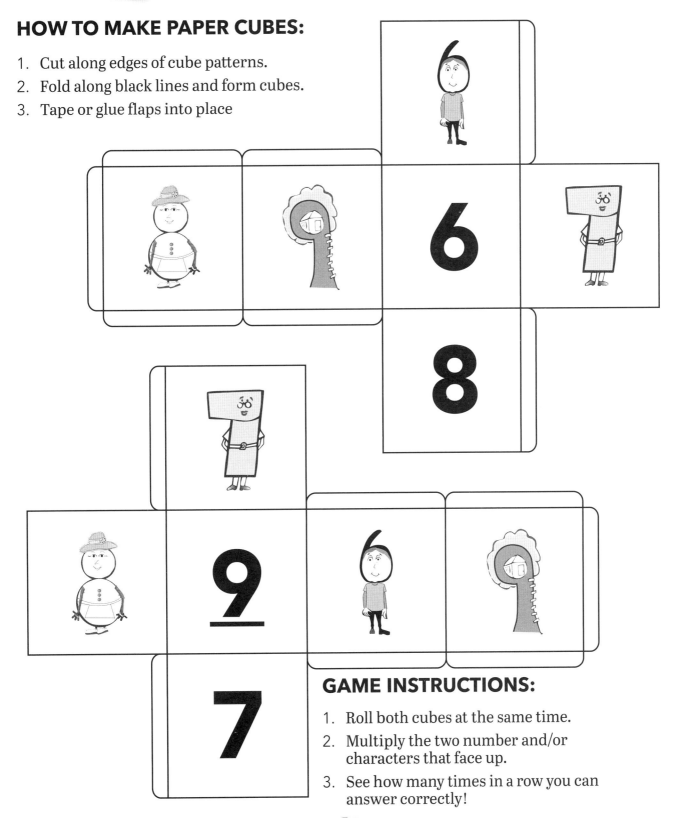

## GAME INSTRUCTIONS:

1. Roll both cubes at the same time.
2. Multiply the two number and/or characters that face up.
3. See how many times in a row you can answer correctly!

Part 1

## MEMORY STORY BOOKLET INSTRUCTIONS:

1. Copy p. 81 & 82 back to back on a single page.
2. Cut along dotted lines.
3. Fold along solid line.
4. Assemble according to page numbers.
5. Staple twice along fold.

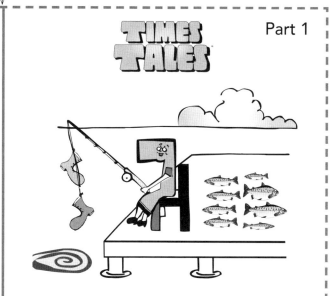

Part 1

You Can Do It!

| | |
|---|---|
| 7 x 3 = | 9 x 3 = |
| 9 x 4 = | 6 x 4 = |
| 3 x 9 = | 7 x 4 = |
| 8 x 4 = | 3 x 6 = |
| 4 x 9 = | 3 x 8 = |
| 6 x 3 = | 4 x 7 = |
| 4 x 6 = | 3 x 7 = |
| 4 x 8 = | 8 x 3 = |

9

Mrs. Week went Butterfly hunting. She captured 20 in her net and 1 landed on her head.

2

The Chair got stuck in the Treehouse after someone tied on 3 kites and 6 balloons.

7

A Butterfly flew by the Treehouse and discovered a treasure of 27 cents.

4

81

**Part 1**  ✂ Cut Along Dotted Lines   Fold

The Sixth Grade Class played musical Chairs for 24 hours.   1

© Copyright 2017 All rights reserved
Published by Trigger Memory Co. LLC   10

Mrs. Snowman stood on a Chair to reach her 3 buttons and 2 mittens.   3

After the warm weather melted Mrs. Snowman, 2 Butterflies took her arms and 4 flew away with her apron.   8

Mrs. Week sat on a Chair and fished. She caught 2 boots and 8 trout.   5

The Sixth Grade Class raised Butterflies. At 1 o'clock, they set 8 Butterflies free.   6

## MEMORY STORY BOOKLET INSTRUCTIONS:

1. Copy p. 83 & 84 back to back on a single page.
2. Cut along dotted lines.
3. Fold along solid line.
4. Assemble according to page numbers.
5. Staple twice along fold.

### You Can Do It!

| | | | |
|---|---|---|---|
| 6 x 6 = | 9 x 8 = | 7 x 9 = | 8 x 6 = |
| 9 x 7 = | 7 x 7 = | 9 x 6 = | 8 x 9 = |
| 9 x 9 = | 8 x 8 = | 6 x 8 = | 7 x 6 = |
| 6 x 9 = | 6 x 7 = | 8 x 7 = | 7 x 8 = |

Part 2

Part 2

Mrs. Snowman went to the Treehouse and knocked 7 times on 2 doors.

9

Mrs. Week and Mrs. Snowman were driving together and went one mile over the speed limit.

2

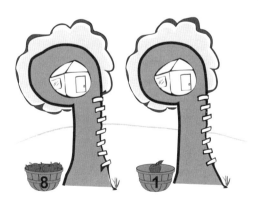

There were two Treehouses. The first Treehouse grew 8 apples and the second grew only 1 apple.

7

Two Sixth Grade Classes played soccer and the score was 3 to 6.

4

83

# Part 2

✂ Cut Along Dotted Lines

Mr. and Mrs. Week have 4 dogs and 9 cats.

1

The Sixth Grade Class went to the Treehouse to feed 5 pounds of bananas to 4 monkeys.

10

Mrs. Week's Sixth Grade Class has 42 students.

3

Mrs. Snowman visited the Sixth Grade Class and showed them 4 birds and an 8 legged spider.

8

Mrs. Week went to the Treehouse and raked up 6 bags of leaves by 3 o'clock.

5

Mr. and Mrs. Snowman must eat 6 snow cones 4 times a day.

6

# Division

DO NOT try this section until Parts 1 & 2 have been mastered.

Part 1

✂ Cut Along Dotted Lines

27 ÷ 🦋 =

What is missing?

24 ÷ 🪑 =

Who is missing?

28 ÷ 🪑 =

Who is missing?

18 ÷ 🧑 =

What is missing?

24 ÷ 👒 =

What is missing?

28 ÷ 📏 =

What is missing?

21 ÷ 📏 =

What is missing?

24 ÷ 🦋 =

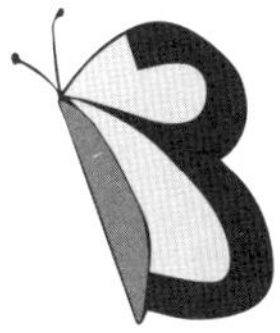

Who is missing?

Part 1 ✂ Cut Along Dotted Lines

| $24 \div 4 =$ | $27 \div 3 =$ |
| --- | --- |
| $18 \div 6 =$ | $28 \div 4 =$ |
| $28 \div 7 =$ | $24 \div 8 =$ |
| $24 \div 3 =$ | $21 \div 7 =$ |

✂ Cut Along Dotted Lines                                                                 Part 1

21 ÷ 🦋 =

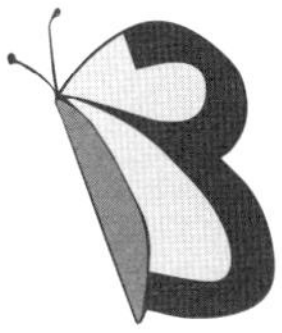

Who is missing?

36 ÷ 🪑 =

What is missing?

27 ÷ 🌳 =

What is missing?

32 ÷ 🪑 =

Who is missing?

36 ÷ 🌳 =

What is missing?

18 ÷ 🦋 =

Who is missing?

24 ÷ 🧒 =

What is missing?

32 ÷ ⛄ =

What is missing?

89

Part 1   ✂ Cut Along Dotted Lines

| 36 ÷ 4 = | 21 ÷ 3 = |
| --- | --- |
| 32 ÷ 4 = | 27 ÷ 9 = |
| 18 ÷ 3 = | 36 ÷ 9 = |
| 32 ÷ 8 = | 24 ÷ 6 = |

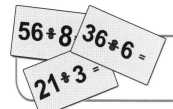

# Division Quiz

Part 1

Name:_____
Time:_____
Score:____

Write the correct answer in each box.

27 ÷ 🦋(3) =
What is missing?

28 ÷ 🪑(4) =
Who is missing?

24 ÷ ⛄(?) =
What is missing?

21 ÷ 👗(7) =
What is missing?

24 ÷ 🪑(4) =
Who is missing?

18 ÷ 👦(6) =
What is missing?

28 ÷ 👗(7) =
What is missing?

24 ÷ 🦋(3) =
Who is missing?

21 ÷ 🦋(3) =
Who is missing?

27 ÷ 👧(?) =
What is missing?

36 ÷ 👧(?) =
What is missing?

24 ÷ 👦(6) =
What is missing?

36 ÷ 🪑(4) =
What is missing?

32 ÷ 🪑(4) =
Who is missing?

18 ÷ 🦋(3) =
Who is missing?

32 ÷ ⛄(?) =
What is missing?

91

# Part 1

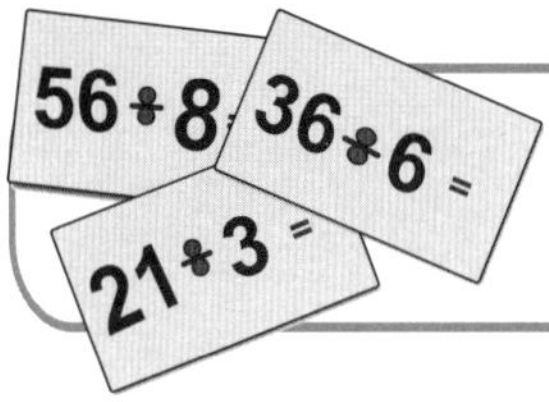

## Division Test

Name:_____
Time:_____
Score:_____

Write the correct answer in each box.

24 ÷ 8 =

27 ÷ 9 =

27 ÷ 3 =

21 ÷ 3 =

32 ÷ 8 =

36 ÷ 9 =

21 ÷ 7 =

24 ÷ 3 =

24 ÷ 4 =

36 ÷ 4 =

18 ÷ 6 =

32 ÷ 4 =

28 ÷ 7 =

18 ÷ 3 =

24 ÷ 6 =

28 ÷ 4 =

Challenge your skills further with the Times Tales® Game Show Quiz DVD

Part 2

✂ Cut Along Dotted Lines

56 ÷ 🂠 =

Who is missing?

56 ÷ 🂠 =

Who is missing?

48 ÷ 🂠 =

Who is missing?

64 ÷ 🂠 =

Who is missing?

42 ÷ 🂠 =

Who is missing?

49 ÷ 🂠 =

Who is missing?

42 ÷ 🂠 =

Who is missing?

36 ÷ 🂠 =

Who is missing?

93

Part 2  ✂ Cut Along Dotted Lines

| 56 ÷ 7 = | 56 ÷ 8 = |
| --- | --- |
| 64 ÷ 8 = | 48 ÷ 6 = |
| 49 ÷ 7 = | 42 ÷ 7 = |
| 36 ÷ 6 = | 42 ÷ 6 = |

Part 2

✂ Cut Along Dotted Lines

54 ÷ 🧑 =

What is missing?

63 ÷ 🧑 =

What is missing?

72 ÷ 🧑 =

What is missing?

81 ÷ 🧑 =

What is missing?

54 ÷ 🧑 =

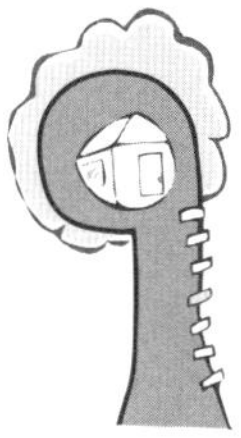

Who is missing?

63 ÷ 🧑 =

Who is missing?

72 ÷ 🧑 =

Who is missing?

48 ÷ 🧑 =

Who is missing?

95

Part 2

| | |
|---|---|
| 63 ÷ 7 = | 54 ÷ 6 = |
| 81 ÷ 9 = | 72 ÷ 8 = |
| 63 ÷ 9 = | 54 ÷ 9 = |
| 48 ÷ 8 = | 72 ÷ 9 = |

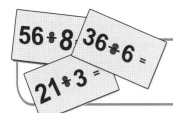

# Division Quiz

Part 2

Name:_____
Time:_____
Score:_____

Write the correct answer in each box.

56 ÷ 🂠 =
Who is missing?

54 ÷ 🂠 =
What is missing?

48 ÷ 🂠 =
Who is missing?

72 ÷ 🂠 =
What is missing?

42 ÷ 🂠 =
Who is missing?

63 ÷ 🂠 =
Who is missing?

56 ÷ 🂠 =
Who is missing?

48 ÷ 🂠 =
Who is missing?

42 ÷ 🂠 =
Who is missing?

63 ÷ 🂠 =
What is missing?

64 ÷ 🂠 =
Who is missing?

81 ÷ 🂠 =
What is missing?

49 ÷ 🂠 =
Who is missing?

54 ÷ 🂠 =
Who is missing?

72 ÷ 🂠 =
Who is missing?

36 ÷ 🂠 =
Who is missing?

# Part 2

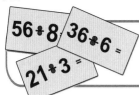

## Division Test

Name:_____
Time:_____
Score:____

Write the correct answer in each box.

| 63 ÷ 7 = | | 54 ÷ 9 = | |
| 48 ÷ 6 = | | 72 ÷ 8 = | |
| 42 ÷ 7 = | | 56 ÷ 7 = | |
| 48 ÷ 8 = | | 72 ÷ 9 = | |
| 42 ÷ 6 = | | 63 ÷ 9 = | |
| 64 ÷ 8 = | | 81 ÷ 9 = | |
| 49 ÷ 7 = | | 56 ÷ 8 = | |
| 54 ÷ 6 = | | 36 ÷ 6 = | |

 Challenge your skills further with the Times Tales® Game Show Quiz DVD

# ANSWER KEY

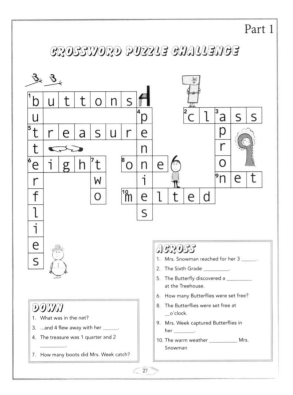

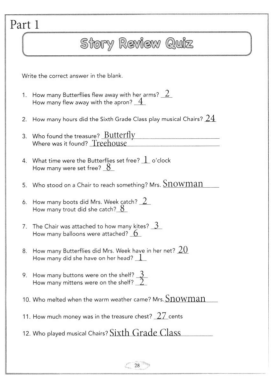

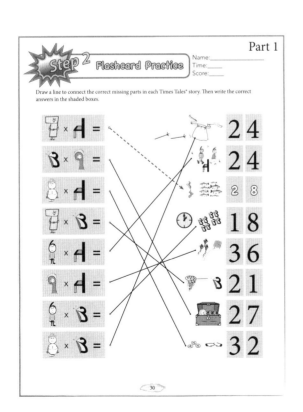

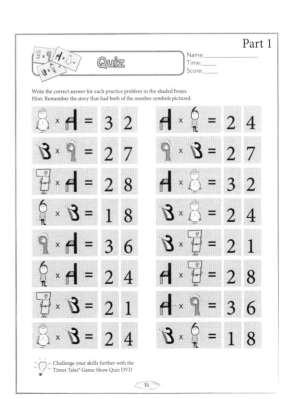

# ANSWER KEY

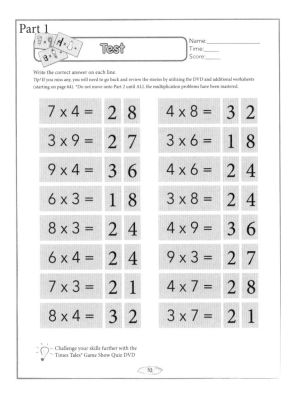

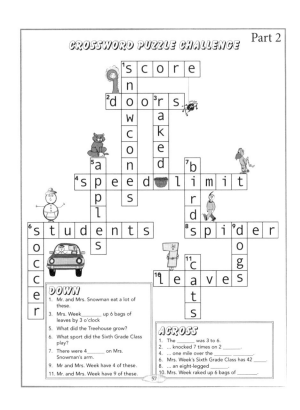

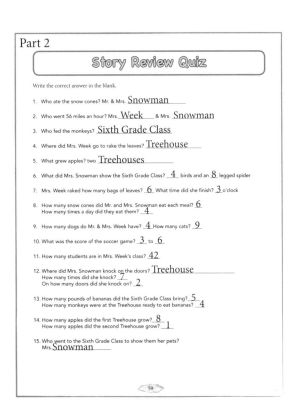

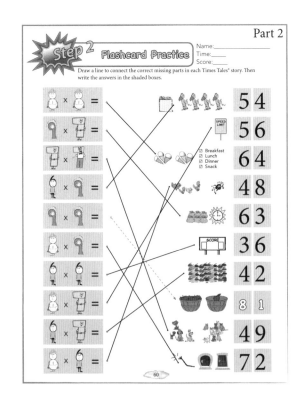

# ANSWER KEY

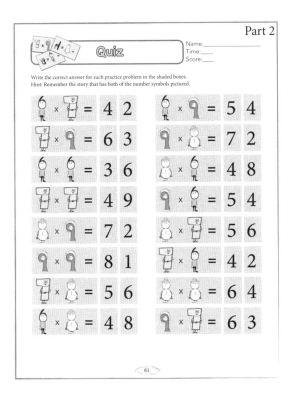

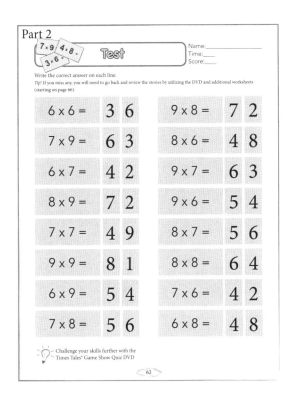

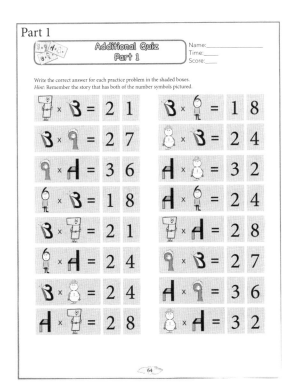

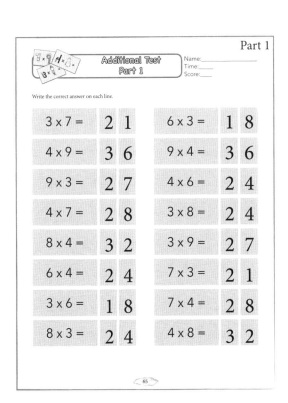

# ANSWER KEY

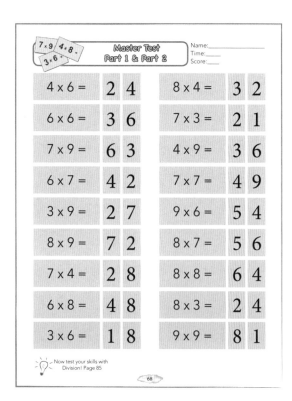

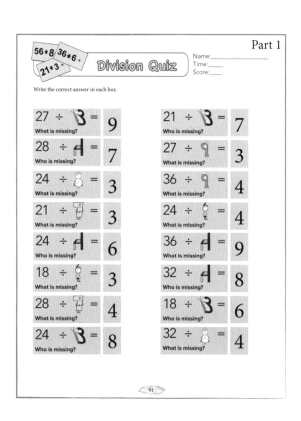

# ANSWER KEY

## Part 1 — Division Test

Write the correct answer in each box.

| | | | |
|---|---|---|---|
| 24 ÷ 8 = | 3 | 27 ÷ 9 = | 3 |
| 27 ÷ 3 = | 9 | 21 ÷ 3 = | 7 |
| 32 ÷ 8 = | 4 | 36 ÷ 9 = | 4 |
| 21 ÷ 7 = | 3 | 24 ÷ 3 = | 8 |
| 24 ÷ 4 = | 6 | 36 ÷ 4 = | 9 |
| 18 ÷ 6 = | 3 | 32 ÷ 4 = | 8 |
| 28 ÷ 7 = | 4 | 18 ÷ 3 = | 6 |
| 24 ÷ 6 = | 4 | 28 ÷ 4 = | 7 |

Challenge your skills further with the Times Tales® Game Show Quiz DVD

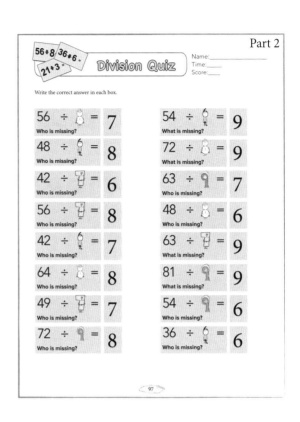

## Part 2 — Division Quiz

Write the correct answer in each box.

| | | | |
|---|---|---|---|
| 56 ÷ ◯ = | 7 | 54 ÷ ◯ = | 9 |
| Who is missing? | | What is missing? | |
| 48 ÷ ◯ = | 8 | 72 ÷ ◯ = | 9 |
| Who is missing? | | What is missing? | |
| 42 ÷ ◯ = | 6 | 63 ÷ ◯ = | 7 |
| Who is missing? | | What is missing? | |
| 56 ÷ ◯ = | 8 | 48 ÷ ◯ = | 6 |
| Who is missing? | | Who is missing? | |
| 42 ÷ ◯ = | 7 | 63 ÷ ◯ = | 9 |
| Who is missing? | | What is missing? | |
| 64 ÷ ◯ = | 8 | 81 ÷ ◯ = | 9 |
| Who is missing? | | What is missing? | |
| 49 ÷ ◯ = | 7 | 54 ÷ ◯ = | 6 |
| Who is missing? | | Who is missing? | |
| 72 ÷ ◯ = | 8 | 36 ÷ ◯ = | 6 |
| Who is missing? | | Who is missing? | |

## Part 2 — Division Test

Write the correct answer in each box.

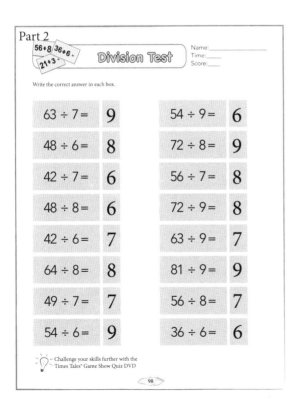

| | | | |
|---|---|---|---|
| 63 ÷ 7 = | 9 | 54 ÷ 9 = | 6 |
| 48 ÷ 6 = | 8 | 72 ÷ 8 = | 9 |
| 42 ÷ 7 = | 6 | 56 ÷ 7 = | 8 |
| 48 ÷ 8 = | 6 | 72 ÷ 9 = | 8 |
| 42 ÷ 6 = | 7 | 63 ÷ 9 = | 7 |
| 64 ÷ 8 = | 8 | 81 ÷ 9 = | 9 |
| 49 ÷ 7 = | 7 | 56 ÷ 8 = | 7 |
| 54 ÷ 6 = | 9 | 36 ÷ 6 = | 6 |

Challenge your skills further with the Times Tales® Game Show Quiz DVD

# WORKBOOK EDITION

Jennie von Eggers
& Marillee Flanagan

© Copyright 2017 All rights reserved
Published by Trigger Memory Co. LLC

*Cover & Interior Artwork Design by Jennie von Eggers*
*Edited by Marillee Flanagan & Janet Haddock*
*Published by Trigger memory Co. LLC*
*ISBN: 978-0-9863000-4-2*

For more information and to see our Classroom Editions of Times Tales® please visit us at **www.TimesTales.com** or contact:
Trigger Memory Co. LLC
2395 Delle Celle Dr., Richland, WA 99354
Email: **TriggerMemory1@gmail.com**
Customer Service: 541-377-0064 Mon.-Fri. 9-5 (PST)

Trigger Memory Co. grants permission to photocopy ONLY the reproducible pages in the *back of this book* (pages 64-98). No other part of this publication may be reproduced in whole or part, transmitted or stored electronically or via any other means without the consent of the publisher.